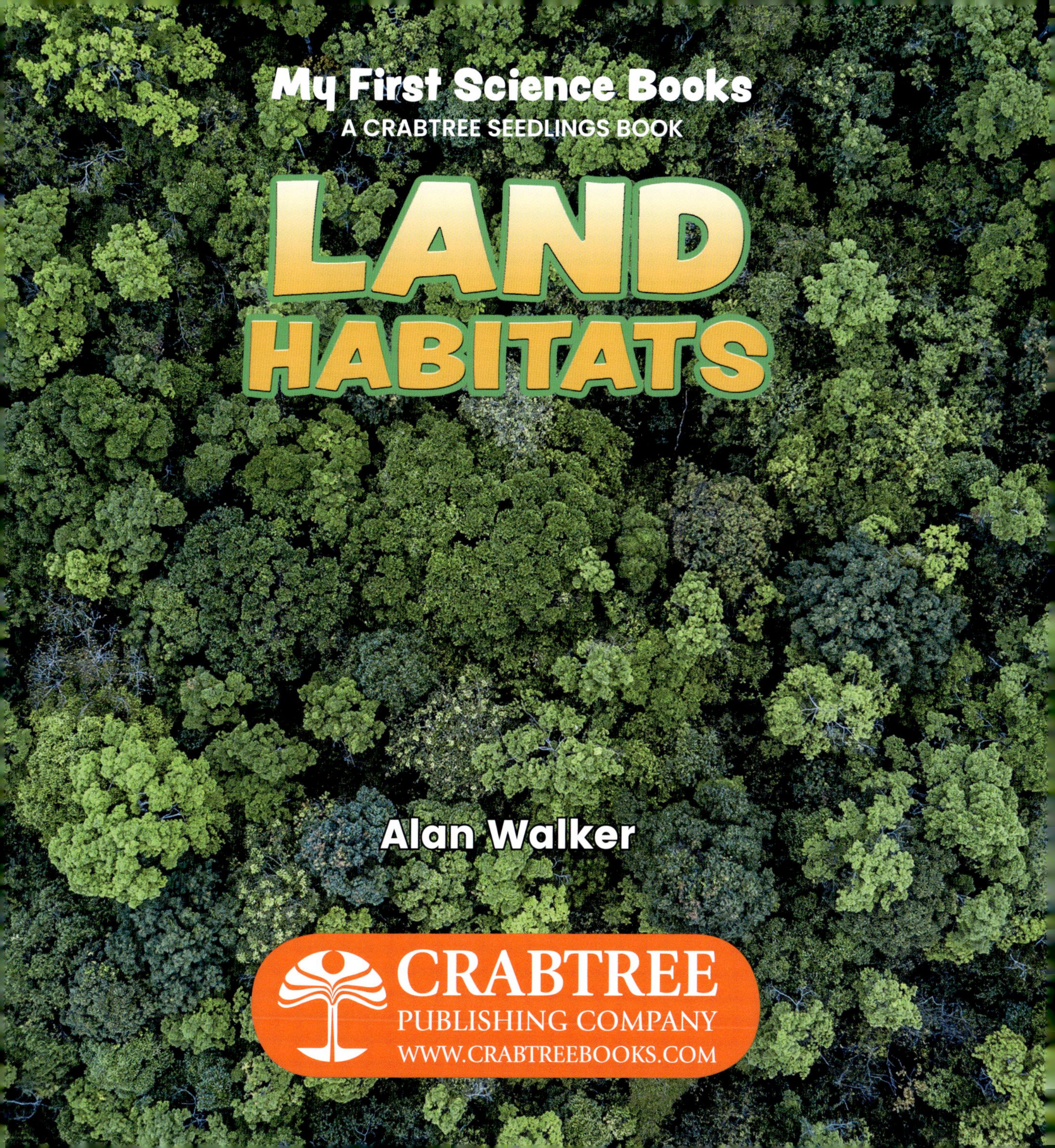

My First Science Books

A CRABTREE SEEDLINGS BOOK

LAND HABITATS

Alan Walker

CRABTREE
PUBLISHING COMPANY
WWW.CRABTREEBOOKS.COM

Different kinds of animals live in different kinds of habitats.

a toucan in a rainforest habitat

a polar bear in an arctic habitat

owl

Owls, raccoons, and deer live in **forest** habitats.

They find **shelter** in the trees and plants.

raccoon

deer

They find food
and water in forests.

Scorpions, rattlesnakes, and rabbits live in **desert** habitats.

scorpion

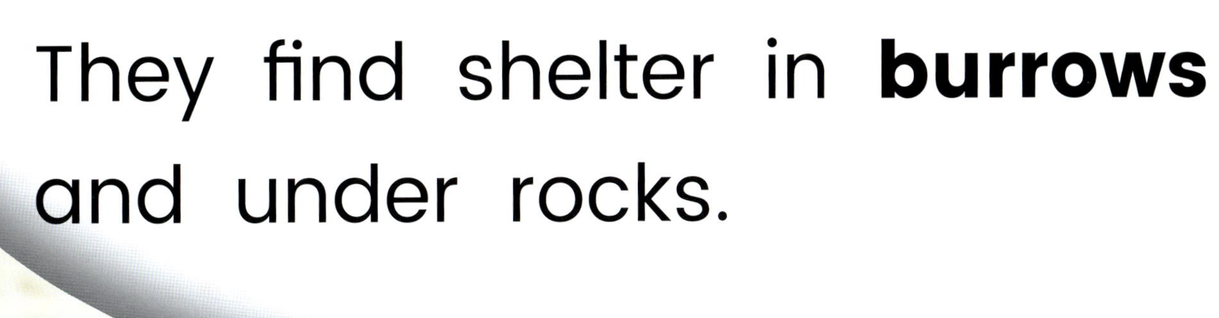

They find shelter in **burrows** and under rocks.

rattlesnake

Deserts are dry, but many desert animals get water from **cactuses.**

desert cottontail rabbit

Sometimes habitats are changed by people.

We cut down forests and **build** in deserts.

When this happens, animals lose their habitats.

Glossary

build (BILD): To build is to make something by putting parts together.

burrows (BUR-ohz): Burrows are tunnels or holes that animals, such as rabbits, use for shelter.

cactuses (KAK-tuhss-iz): Cactuses are plants with thick trunks and spikes. They grow in hot, dry areas.

desert (DEZ-ert): A desert is a dry area that gets very little rain.

forest (FOR-ist): A forest is a large area thickly covered with trees.

shelter (SHEL-tur): Shelter is a place where an animal can live and hide from bad weather or danger.

Index

School-to-Home Support for Caregivers and Teachers

Crabtree Seedlings books help children grow by letting them practice reading. Here are a few guiding questions to help the reader with building his or her comprehension skills. Possible answers are included.

Before Reading

- **What do I think this book is about?** I think this book is about habitats on land. A habitat is a place where a plant or animal lives.
- **What do I want to learn about this topic?** I want to learn about the different plants and animals found in land habitats.

During Reading

- **I wonder why...** I wonder why people change habitats.
- **What have I learned so far?** I have learned that trees, owls, raccoons, and deer live in forest habitats. I have learned that cactuses, scorpions, rattlesnakes, and rabbits live in desert habitats.

After Reading

- **What details did I learn about this topic?** I learned that desert animals get water from cactuses.
- **Read the book again and look for the vocabulary words.** I see the word ***shelter*** on page 6 and the word ***burrows*** on page 13. The other vocabulary words are found on pages 22 and 23.

Library and Archives Canada Cataloguing in Publication

Title: Land habitats / Alan Walker.
Names: Walker, Alan, 1963- author.
Description: Series statement: My first science books | "A Crabtree seedlings book". | Includes index. | Previously published in electronic format by Blue Door Education in 2020.
Identifiers: Canadiana 20200389521 | ISBN 9781427130266 (hardcover) | ISBN 9781427130372 (softcover)
Subjects: LCSH: Habitat (Ecology)—Juvenile literature.
Classification: LCC QH541.14 .W35 2021 | DDC j577—dc23

Library of Congress Cataloging-in-Publication Data

Names: Walker, Alan, 1963- author.
Title: Land habitats / Alan Walker.
Description: New York : Crabtree Publishing, 2021. | Series: My first science books : a Crabtree seedlings book | Includes index.
Identifiers: LCCN 2020051012 | ISBN 9781427130266 (hardcover) | ISBN 9781427130372 (paperback)
Subjects: LCSH: Habitat (Ecology)--Juvenile literature. | Habitat conservation--Juvenile literature.
Classification: LCC QH541.14 .W33 2021 | DDC 577--dc23
LC record available at https://lccn.loc.gov/2020051012

Crabtree Publishing Company
www.crabtreebooks.com 1–800–387–7650

Author: Alan Walker
Production coordinator and Prepress technician: Tammy McGarr
Print coordinator: Katherine Berti

e-book ISBN 978-1-949354-36-2

Print book version produced jointly with Blue Door Education in 2021

Printed in the U.S.A./012021/CG20201112

Photo credits: Cover: ©shutterstock.com/Erik Mandre, ground squirrels © shutterstock.com/David G Hayes, icon top right © shutterstock.com/baza178, page 4 ©shutterstock.com/ Fedor Selivanov, page 5 © shutterstock.com/Alexey Seafarer, page 6-7 © shutterstock.com/Pecak, page 8-9 © shutterstock.com/Tony Campbell, page 10-11 © shutterstock.com/Dark Moon Pictures, page 12-13 © shutterstock.com/IrinaK, page 14-15 © shutterstock.com/Mark_Kostich, page 16-17 © shutterstock.com/Charles T. Peden, page 18-19 © shutterstock.com/Timelynx, page 20-21 © shutterstock.com/Heder Zambrano, page 22-23 © shutterstock.com/Jay Ondreicka, page 24: top photo © shutterstock.com/Joe Mercier, middle photo © shutterstock.com/Lindasj22, bottom photo © shutterstock.com/Sibella Bombal, page 25: top photo © shutterstock.com/Enate Images, middle photo © shutterstock.com/attila

Published in Canada
Crabtree Publishing
616 Welland Ave.
St. Catharines, Ontario
L2M 5V6

Published in the United States
Crabtree Publishing
347 Fifth Ave.
Suite 1402-145
New York, NY 10016

Published in the United Kingdom
Crabtree Publishing
Maritime House
Basin Road North, Hove
BN41 1WR

Published in Australia
Crabtree Publishing
Unit 3 – 5 Currumbin Court
Capalaba
QLD 4157